THE SCIENCE DETECTIVE INVESTIGATES

Materials

Katie Dicker

WAYLAND

First published in 2010 by Wayland

Copyright © Wayland 2010

Wayland
338 Euston Road
London NW1 3BH

Wayland Australia
Level 17/207 Kent Street
Sydney, NSW 2000

Produced for Wayland by
White-Thomson Publishing Ltd

www.wtpub.co.uk
+44 (0)845 362 8240

Senior editor: Camilla Lloyd
Designer: Clare Nicholas
Consultant: Jon Turney
Picture researcher: Amy Sparks
Illustrator: Stefan Chabluk
Sherlock Bones artwork: Richard Hook

British Library Cataloguing in Publication Data:
Dicker, Katie.
    Materials. – (The science detective investigates)
    1. Materials–Juvenile literature. 2. Materials–
Experiments–Juvenile literature.
    I. Title II. Series
    620.1'1-dc22

ISBN 978 0 7502 6018 3

Printed in China

Wayland is a division of Hachette Children's Books, an Hachette UK company.

www.hachette.co.uk

Picture Acknowledgments:
Abbreviations: t-top, b-bottom, l-left, r-right, m-middle.
Cover: Dreamstime (Lagui)
Insides: Folios Istockphoto (Macroworld); **1** Dreamstime (Monkey business); **4** Dreamstime (l Phartisan, r Miguelpinheiro); **6** (l) Dreamstime (Rick Lord), (r) Getty Images (Ryan McVay); **8** Dreamstime (Gonfig); **9** (l) Istockphoto (Gabor Izso), (r) Dreamstime (Elena Elisseeva); **11** Istockphoto (Tim MacPherson); **12** (t) Istockphoto (iofoto), (b) Dreamstime (Canbalci); **13** (t) Dreamstime (Grandpa), (b) Photolibrary (jutta klee), (r) Shutterstock (Russ Du parcq); **15** (t) Istockphoto (TayaCho), (b) Dreamstime; **16** Dreamstime (t Odusan, b Irochka); **17** Dreamstime (Carrie Anne); **18** Dreamstime (l Monkey business, mr Robynmac); **19** (l) Dreamstime (Lagui), (mr) Shutterstock (immelstorm); **20** Shutterstock (graph); **21** Dreamstime (Ukapala); **22** (t) Dreamstime (Photowitch), (b) Getty Images (Image Source); **23** Dreamstime (Nolexa); **24** Dreamstime (t Gemenacom, b Tinabelle); **26** Dreamstime (Domhnall); **27** Photolibrary (Japan Travel Bureau); **28** Getty Images (Justin Bailie); **29** Getty Images (G. Wanner).

Contents

Words that appear in **bold** can be
found in the glossary on page 30.

**The Science Detective, Sherlock Bones, will help you learn all about
Materials. The answers to Sherlock's questions can be found on page 31.**

What are materials?

Look around and you will see a world full of materials. All the objects and items that we use in our daily lives are made from materials – from wooden pencils and plastic pens to metal mugs and china cups. We use materials to build our homes, to make our clothes and to carry out everyday tasks.

Natural materials

Some materials are found naturally on Earth. Wood comes from trees, stones are found in the ground and wool is taken from a sheep's fleece (coat). Sometimes, natural materials are **processed** to improve them. Diamonds are shaped and polished to make them sparkle and shine. Wood can be cut and carved into shapes.

Man-made materials

Over thousands of years, different materials have been tried and tested. Many of the materials we use today are not found naturally on Earth. They are made in special chemical processes that change or combine natural materials. Plastic and glass are two examples. Plastic can be **moulded** to make boxes, bottles and bin bags. Plastic is made from chemicals found in oil deep underground. Glass is used to make windows, drinking glasses and windscreens for cars. Sand and **limestone** are two natural **substances** used to make glass.

▼ **Can you identify some of the materials used to build this house in the USA? What materials have been used to build this home in Africa (right)? Why do you think these materials differ?**

🐾 **Use books or the Internet to find out how some man-made materials (such as glass, paper and plastic) are made from natural materials.**

THE SCIENCE DETECTIVE INVESTIGATES:

Natural or man-made?

Look at the following illustrations of common materials and identify which materials are natural and which are man-made. Make a table to record your findings. Are any of the materials natural but processed?

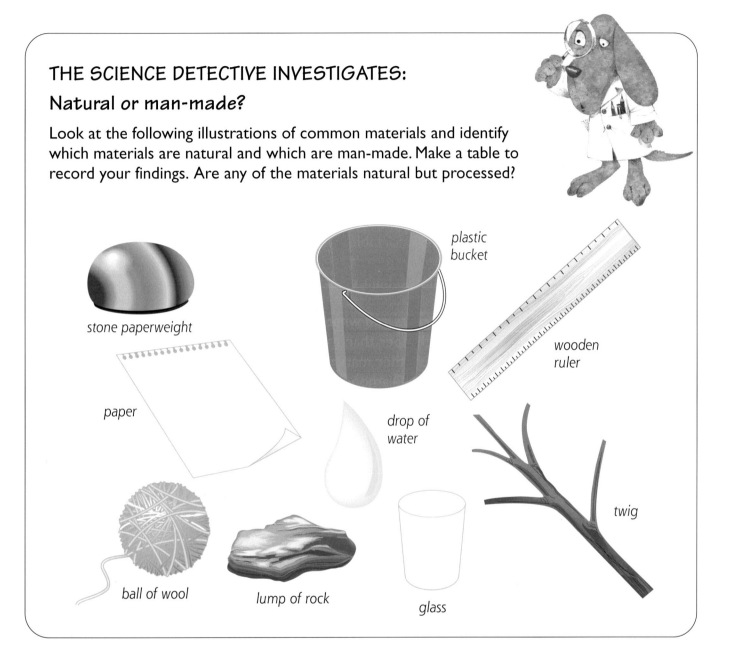

stone paperweight

plastic bucket

wooden ruler

paper

drop of water

twig

ball of wool

lump of rock

glass

SCIENCE AT WORK

Scientists are constantly looking at the **properties** of natural materials to help them create new materials of their own. Spider silk, for example, is one of the strongest natural materials known to man. It is ten times thinner than a human hair but stronger than a steel wire of the same thickness. Spider silk is also **biodegradable**. It could be used to make strong, light protective clothing or strong, fine surgical thread. Scientists in the USA have been looking at ways to manufacture spider silk on a large scale.

How do we use materials?

Materials have different properties depending on the substances they are made from, and the way these substances are combined. We choose materials to suit the jobs we want them to do. A plastic cup is light and doesn't break when you drop it. Transparent glass lets sunlight through a window. A woollen jumper keeps you warm on a cold day.

Choosing materials

When we choose which materials to use, we look at a material's properties. A springy mattress makes a comfortable bed to lie on and a strong brick wall stays standing on a windy day — imagine lying on a metal mattress or living in a house built from paper!

We also consider how much a material costs and whether it is easy to find. Local materials are usually cheaper than materials that are transported from place to place. Other materials are expensive because they are difficult to find. Diamond, for example, is **mined** from rocks deep underground. Diamonds are cut and polished to make sparkling jewellery. They are also used in industry to cut, grind and polish other surfaces because their hard surfaces don't wear away. But diamond is too expensive to use as an everyday material.

▼ **On a wet day, we wear waterproof clothes and use an umbrella to keep ourselves dry.**

▼ **Diamond is the hardest material on Earth. It is used in industry and to make expensive jewellery.**

THE SCIENCE DETECTIVE INVESTIGATES:

Testing properties

1 Collect some everyday items made from metal, wood, plastic and fabric. Add a piece of modelling clay. Try to choose some examples made from the same material (such as a metal spoon and a metal paper clip).

2 Feel each item in turn and describe its properties. What does each item look and smell like? What sound does it make when you tap it gently against a table?

3 Test your materials in a bowl of water to see if they float or sink. Are they waterproof or absorbent?

4 Record your results in a table. Then draw a **Venn diagram** to help you compare some of the similarities and differences of each material. Use the example (right) as a guide.

🐾 Look at this list of properties. Can you think of objects around you that have these properties? When and why are these properties useful? What other common properties can you think of?

soft

smooth

sharp

shiny

transparent

light

absorbent

waterproof

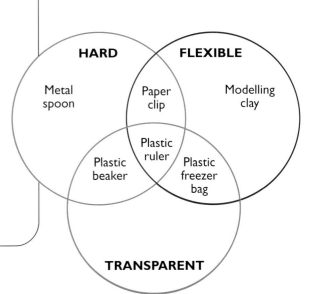

What are solids and liquids?

The materials around us come in three different forms: **solids**, **liquids** and **gases**. We call these forms a material's 'state'. A solid is a material that can keep its shape and volume. This book is solid and so is the chair you are sitting on. A liquid is a material that keeps its volume but takes the shape of the container it is in, such as water filling a glass or a vase.

Solid shapes

Solids come in different shapes and sizes. They can be heavy or light and hard or soft. A dry bar of soap, for example, feels hard to touch and is quite heavy. A bath sponge of the same size is soft to touch. It is also light to carry because the sponge is full of air pockets. Some solids, such as wood and clay, can be cut or moulded to change their shape.

◀ **This chair is a solid. It holds its shape when you sit on it.**

SCIENCE AT WORK

Some solids behave like liquids. Sand, for example, is a solid but it can be poured just like a liquid. Although the tiny grains of sand hold their individual shapes, when the grains are grouped together, they move around so the sand takes the shape of the container it is in.

🐾 **How can you cause a solid to change its shape?**

Flowing liquids

Water and honey are both liquids. They spread out or splash if you spill them. Liquids flow and allow solids to pass through them. If you throw a pebble into a pond, for example, it sinks through the water until it reaches a solid surface. Liquids change shape but they don't change their volume. Water is a thin liquid that pours quickly and splashes easily. Honey is a thick, sticky liquid that pours more slowly.

▶ **When you walk through sea water, it splashes against you.**

▼ **A wooden dipper is used to pick up runny honey.**

What are gases?

Gases are usually difficult to see, smell or feel but they are just as important as solids and liquids. Air is a mixture of gases. We need to breathe air to survive. Unlike solids and liquids, gases have no shape of their own. They constantly move around and fill whatever space they are in.

Moving gases

Gases flow more easily than liquids. They quickly move to fill the shape of the container they are in. Gases change shape but, unlike liquids, they also change volume. If a gas is released into a room, it spreads out and fills the room. Although gases are difficult to see, they are very common in our daily lives. If you fly a kite, the wind (moving gases in the air) makes it swirl and swoop. When you blow up a balloon, the air in your lungs fills the balloon to make it grow. In a storm, you can feel the strength of the wind blowing against you.

THE SCIENCE DETECTIVE INVESTIGATES:

Proving the presence of air

You will need:
- clear plastic cup • paper towel
- transparent bowl/tank of water (deeper than the cup)

Try this magic trick on your friends!

1 Push a paper towel firmly into the bottom of a clear plastic cup.
2 Turn the cup upside down and lower it (vertically) into a transparent tank of water until the cup is completely submerged.
3 Lift the cup out (vertically) and check the paper towel. Is it dry? The paper towel stays dry because air in the cup stops the water from reaching it.
4 Repeat the experiment again and look at the water level. Why does a small amount of water rise in the cup? What happens when you tilt the cup?

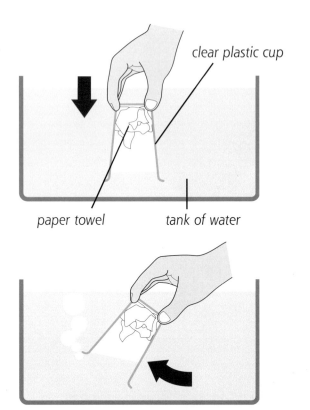

clear plastic cup

paper towel   tank of water

SCIENCE AT WORK

Gases change their shape and volume by filling the container they are in. But gases can also be **compressed**. This makes it easier to store and transport them. Gases also **expand** when they are heated. A hot air balloon rises in the sky because the warm air inside it spreads out and becomes lighter than the air outside.

Use books or the Internet to find out about different gases and how they are used. For example, you could find out how carbon dioxide gas is used to make fizzy drinks or how helium gas is used in balloons.

▼ Moving gases in the air make it windy and blustery. These boats are catching the wind in their sails to move along.

What is evaporation?

Sometimes, liquids turn into gases. This is called **evaporation**. The process of evaporation is important in our lives. It helps wet paint and our washing to dry. When we sweat, water evaporates from our skin to cool us down. We call the process of evaporation a 'change of state'.

▲ Evaporation will help this paint to dry.

Speed of evaporation

Liquids evaporate all the time. If you leave a glass of water for a few months, it will eventually evaporate into a gas called **water vapour** so the glass becomes dry. At low **temperatures**, evaporation is very slow. When water is heated, the speed of evaporation increases and the water eventually **boils**. Water boils at 100°C (212°F).

Types of evaporation

Evaporation is difficult to see but we can observe its effects around us. When puddles dry in the sunshine, the water doesn't disappear completely. As the water is warmed, it changes to water vapour and rises into the air. When we hang wet washing outside, the process of evaporation helps the clothes to dry.

▶ The water in this kettle evaporates into water vapour when it is heated.

In winter, our washing dries slowly because the air around it is cold. In summer, the warmer weather causes the water to evaporate at a faster rate. On a warm, windy day, the clothes dry even faster – as the wind moves the water vapour away, there is room for more water to evaporate. Gaps in some fabrics help air to pass through them which speeds up the drying process even more.

▲ **This washing will dry faster on a warm, windy day. Hanging the washing to increase the surface area exposed to the air will make it dry even quicker.**

✿ Why do you think we smell paint across a room?

SCIENCE AT WORK

Evaporation is useful for drying but it has other uses, too. Sometimes, evaporation is used to make concentrated orange juice. The juice is heated so that about 85 per cent of the water evaporates. This reduces it to about one sixth of its volume, making the juice easier to store and transport. The concentrated juice can then be **diluted** with water to make orange juice.

◄▼ **Evaporation can help to make the juice from oranges (left) easier to store and transport (below).**

What is condensation?

Condensation is the opposite of evaporation. It happens when a gas turns into a liquid. We say that evaporation and condensation are **reversible** changes of state. If you've ever breathed onto a pane of glass, you will have seen the process of condensation at work.

Seeing condensation

The air around us is full of water vapour (a gas). When the water vapour cools, it condenses into liquid water. If you fill a hot bath, some of the water evaporates, filling the room with water vapour. When this water vapour hits a cold surface, such as a window or the bathroom mirror, it cools and condenses forming a thin layer of water. In winter, condensation forms on windows because the glass is much colder than the air in the room.

❧ **Why does condensation form on a drinks can when you take it out of a cold fridge?**

THE SCIENCE DETECTIVE INVESTIGATES:

Condensation in action

You will need:
• small drinking glass • large, clear bowl of hand-hot salted water
• plastic food wrap • ice cubes

1. Put the drinking glass in the middle of the bowl of salted water.
2. Cover the bowl with plastic food wrap and leave for a few hours. What do you notice on the underside of the food wrap? Where does the water come from? Where does it go? What is different about the water in the glass?
3. Repeat, this time placing some ice cubes on top of the food wrap. What do you notice about the size of the water droplets? Where are the water droplets largest?

food wrap

glass bowl

hand-hot salted water

drinking glass

ice cubes

Natural water

Condensation happens in nature, too. Sometimes, **dew** forms on the grass in the early morning or evening. This is because water vapour in the air is cooled on the cold ground and turns into liquid water. Dew is very useful. It gives plants and insects a regular source of water even when there is no rain.

✿ **Why does dew form on the grass in the early morning or evening, but not during the day?**

▼ **Water vapour has cooled and condensed on this spider's web to form tiny droplets of water.**

▲ **Condensation on a window forms trickles of water when you touch it.**

When do materials melt and freeze?

Another reversible change of state occurs when materials **melt** or **freeze**. Liquids can change to a solid when they cool. We call this freezing or **solidifying**. Solids can be changed to a liquid when they are heated. We call this melting.

Melting

Different solids melt at different temperatures. If you leave butter or chocolate out on a warm day, it softens as it begins to melt. If you heat butter or chocolate in a saucepan it melts completely, forming a thin liquid. Other materials have to be heated to a very high temperature before they melt.

Freezing and solidifying

When melted butter is cooled, it turns back into a solid. This is called solidifying. Freezing is the same as solidifying. We say a liquid freezes when it becomes a solid at a very cold temperature. Liquids solidify when they become a solid at an everyday temperature. We say that water freezes at 0°C (32°F) to make ice, for example, while butter solidifies at room temperature.

▲ **Warm chocolate melts into a runny liquid.**

◀ **When wax is heated, it melts into a liquid. Hot wax runs down the side of this candle but as it cools, it solidifies to become solid again.**

🐾 **Why do you think some factories heat solid metal and glass?**

Freezing temperatures

Different liquids freeze at different temperatures. The speed at which a liquid freezes also depends on its surroundings. The water in an ice cube tray will become solid in a freezer but not in a fridge, for example. The time the ice cubes take to freeze depends on the temperature of the freezer, the size of the freezer, the size of the ice cubes and the temperature and **purity** of the water. Pure water freezes more quickly than salt water, for example.

SCIENCE AT WORK

Polar bears live on the ice-covered Arctic Ocean. They hunt seals on the ice and in the chilly waters. But as we burn more fuel for energy, **global warming** is causing temperatures on Earth to rise and the ice is melting. Polar bears are losing their homes and their hunting grounds. With less food to eat, their lives are at risk.

▼ **As temperatures on Earth are rising and the ice caps melt, this polar bear is losing his home.**

How does heat change materials?

When some materials are heated, they change permanently. We call this an **irreversible** change of state. The change is permanent because the heat causes a **chemical reaction** and new substances are produced. When you cook your food, for example, chemical reactions take place. This can change the flavour of your food and make it safe to eat.

Baking

When you bake a cake or make a loaf of bread, the heat from the oven changes the ingredients you are using. Soft, runny cake mixture or thick, sticky dough turns brown and becomes hard and spongy. This change is permanent. It is not possible to change the cake or bread back to cake mixture or dough. When a clay pot is fired, it is baked in a very hot oven. A similar reaction takes place – the soft, slimy clay becomes firm and hard. This change is permanent.

◀ Cooking is an example of an irreversible change of state. When you put cake mix into the oven, the heat changes the cake mixture permanently.

🐾 Use books or the Internet to find out how different types of toffee are made. What is special about honeycomb (sponge toffee)?

▲ When a clay pot is fired, a chemical reaction takes place and the clay changes (right). The pot dries out and become firm. This change is permanent.

Burning

When some materials become very hot, they catch fire and burn. **Burning** is another irreversible change of state that produces new substances. When you toast bread, it begins to burn. It gets hot, it turns brown and then black, and it releases smoke. If the bread becomes too hot it can catch fire and you see a flame. You can never get the original bread back once you have toasted it.

What are mixtures and solutions?

Solids, liquids and gases can be combined to form **mixtures**. When some solids are mixed with liquids they **dissolve** to form a **solution**. Solid mixtures can be seen all around us. Sand is a mixture of tiny rocks, stones and broken shells. Soil is a mixture of mud, bark and leaves. The muesli we eat is a mixture of oats, nuts and fruit.

Mixing solids and liquids

When solids are mixed with liquids, two things can happen. Some solids dissolve to form a solution. When you add sugar to tea, for example, the sugar dissolves to form a sweet liquid. Salt dissolves in water in a similar way. You can't see the sugar or salt but you can taste them.

Other solids do not dissolve in liquids. Sand and chalk sink to the bottom of a liquid mixture instead of dissolving. Remember you should never taste liquids for sugar or salt if you don't know what the liquid contains. The liquid may be harmful so you should ask an adult before you taste anything.

🐾 **After a swim in the sea, when your skin has dried, what are the white marks on your skin?**

▲ **Sand is a mixture of rocks, stones and shells that have been ground to fine grains over thousands of years.**

Fast or slow?

The speed at which a solid dissolves in a liquid can be changed. Grains of sugar dissolve faster than a sugar lump because they have a greater surface area. This means that more water comes into contact with them. Stirring helps the sugar to dissolve more quickly because small particles (pieces) of sugar break off. More sugar dissolves to take the place of the moving solution.

Most mixtures and solutions are reversible changes of state. Solids that dissolve can be separated by evaporation. Other solids can be separated by **filtering** (sieving).

▶ **Stirring your coffee helps to speed up the rate at which the sugar dissolves.**

THE SCIENCE DETECTIVE INVESTIGATES:

Separating materials

You will need:
- jar of water • shells, sand and salt
- spare beakers • small bowl • sieve
- filter funnel and filter paper

Mix the water, shells, sand and salt together in the jar. What does the mixture look like? How would you separate the shells or the sand or the salt from the mixture? What apparatus would you choose? Why can't you extract the salt by filtering?

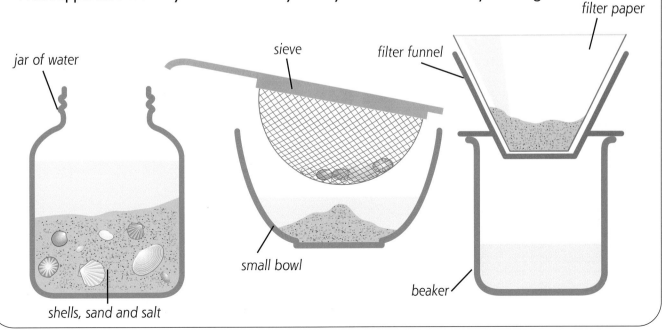

jar of water

sieve

filter funnel

filter paper

small bowl

beaker

shells, sand and salt

Which materials can be squashed and stretched?

Some solids change their shape if you squash, bend, twist or stretch them. Modelling clay is a useful material for making things. If you squash and stretch the clay, it keeps its new shape until you change it again. Other materials, such as springs or rubber balls, snap back to their original shape if you squash or stretch them.

Elasticity

Materials that keep their shape after you stretch them are said to have **elasticity**. An elastic band is useful for keeping a bunch of pencils together – it can be stretched but springs back to its original shape when you let go. A soft rubber ball squashes slightly when you catch it, so it doesn't hurt your hands. Socks are elastic. They stretch when you put them on, to fit the shape of your foot.

◀ The springs on a trampoline have elasticity. When you jump on the material, the springs stretch. But they quickly snap back to their original shape again, helping you to jump even higher.

🐾 **How many materials can you think of that stretch?**

Moulding

Clay is a good modelling material because it can be moulded into lots of different shapes. Some solids, such as glass and plastic, can be moulded when you melt them. When they are soft, these materials can be pulled and stretched. When they cool, they solidify into their new shape.

▶ **When glass is heated, it can be stretched and moulded into new shapes.**

THE SCIENCE DETECTIVE INVESTIGATES:

Hang and stretch

You will need:
- selection of old socks
- washing line and pegs
- selection of marbles (of the same size)

1 Hang a selection of socks on the washing line.
2 Measure each sock (from the line to the toe) and record your measurements in a table.
3 Add the same amount of marbles to each sock and measure them again.
4 Remove the marbles and measure once more. Which socks stretched the furthest? Which socks went back to their original size? Record your results in a bar chart showing the before and after measurements. How would you ensure that your comparisons are fair?

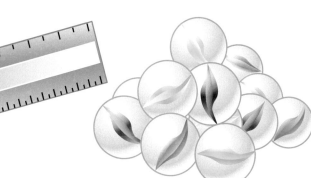

What are conductors and insulators?

Conductors are materials that allow heat (or **electricity**) to pass through them easily. They are useful when we want to warm our food or heat our homes. An **insulator** is a material that does not allow heat (or electricity) to pass through it. Plastic, glass, wood and rubber are all insulators.

Metals

Metals are good conductors of heat. They conduct heat better than wood, plastic, glass and many other non-metals. Copper and aluminium are particularly good conductors. They are often used to make pots and pans. The heat travels through the metals quickly to heat up our food.

Insulators

Sometimes, we don't want materials to get hot. A metal saucepan is good for cooking, but a hot metal handle could burn your hands. Materials, such as plastic, are often used as insulators to keep us safe. A saucepan has a plastic handle to keep it cool. Insulators are also used to keep heat in, as well as keep heat out. In our homes, we use insulating materials, such as foam and glass fibre sheets, in the walls and roof to stop heat escaping to the air outside.

▼ **These glass fibre sheets will help to keep warmth inside a home.**

STAY SAFE

You should take care when touching metals that are near to a source of heat. Saucepans, for example, get hot very quickly and could burn you.

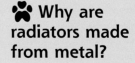 **Why are radiators made from metal?**

THE SCIENCE DETECTIVE INVESTIGATES:

Measuring heat loss

You will need:
• 4 plastic cups with lids, filled with hand-hot water • 4 thermometers
• 4 shoe boxes, with lids • 3 insulating materials such as paper towels, newspaper and cotton wool

1 Put each cup into a shoe box and surround the cup with one of the insulating materials. Leave one cup without any insulation.
2 Ask an adult to help you make a hole in the centre of the shoe box lid and the centre of the cup lid, for the thermometer to go through. Predict which materials you think will be the best insulators.
3 Read the temperature on each thermometer every 10 minutes and record your findings. Plot a graph of your results. The steeper the line falls, the less insulating the material is. How accurate were your predictions? How would you ensure that your comparisons are fair?

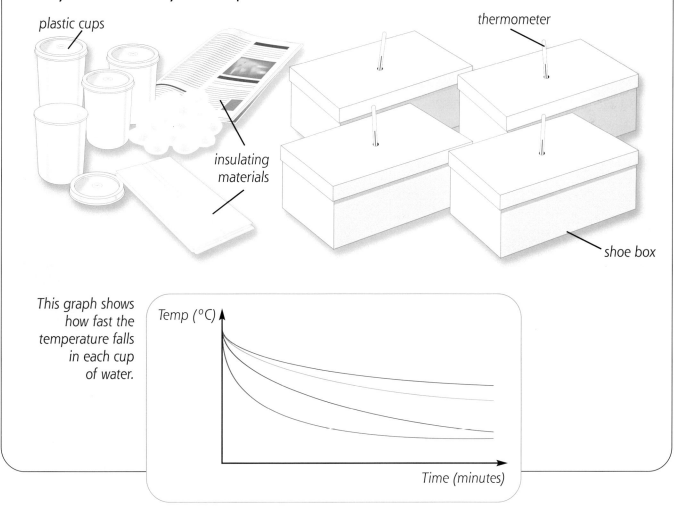

plastic cups

insulating materials

thermometer

shoe box

This graph shows how fast the temperature falls in each cup of water.

Temp (°C)

Time (minutes)

What will materials be like in the future?

Scientists are constantly testing and changing materials to see how they can be used or how they can be improved. Racing cars, for example, used to be made from metal. Today, they are made from **carbon fibre** – a material that is much stronger and lighter than metal. This modern material helps the cars to go even faster.

Modern materials

Scientists have made another very strong, light material called Kevlar®. Made from a type of plastic, Kevlar's lightness and strength makes it a good material for bicycle tyres (to prevent punctures) and as body armour for soldiers. Different types of glues, called **adhesives**, have also been developed in recent years, to make them tougher and stronger, with quicker drying times.

SCIENCE AT WORK

Scientists have produced new materials called 'smart materials'. They respond to changes around them, such as changes in temperature or light. Some spectacles, for example, have special lenses that darken in bright sunlight but stay clear in normal conditions. Memory metals are another example. A memory metal 'remembers' its shape. It is a good material for spectacle frames that might get bent or twisted.

▼ **This soldier's helmet is made from Kevlar. It is light, which allows the soldier to move around easily, but strong for protection.**

Recycling and re-using

Today, we are using more materials than ever before. If we keep throwing materials away, soon there will be none left. **Recycling** helps to save materials. When materials are recycled they are broken down and re-used to make new materials. Many materials, such as paper, plastic, glass and metal, can be recycled. Most of the packaging used in shops today, for example, is made from recycled materials. If we re-use or recycle materials they will last much longer.

🐾 **Use the Internet to find your local recycling centre. What materials can you recycle there?**

▼ **We can help to save materials by recycling them.**

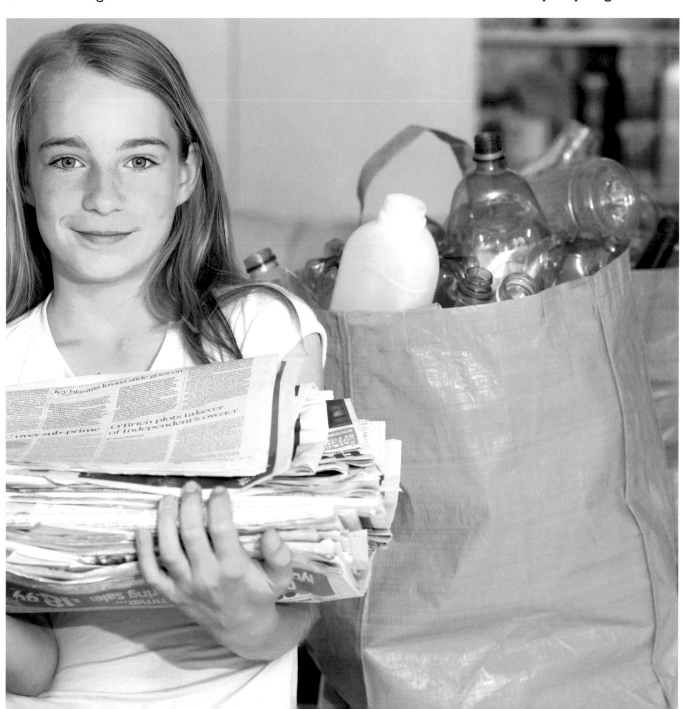

Your project: Testing materials

Imagine you're a scientist developing new materials. What materials would you design to improve the objects that we use? What properties would they need? How would you test them? How would you ensure that your comparisons are fair?

▼ The materials used for ski wear are designed to be comfortable, warm and waterproof in the cold, wet snow.

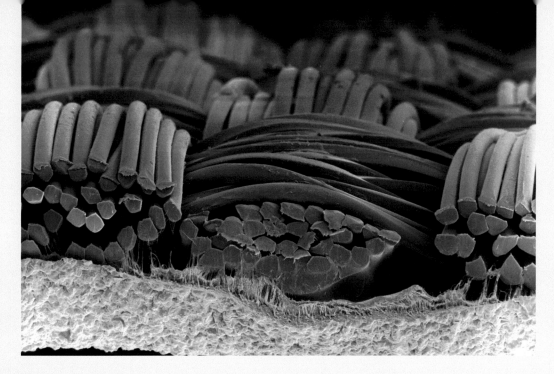

◄ **This highly magnified image shows the layers of material that make Gore-Tex – a waterproof, breathable fabric.**

Materials of the future

Choose one of the following examples (or think of an idea of your own) to design and test materials for a new product. Use the real-life example in the panel below for inspiration. Write a description of the materials you would like to make and think of ways that you could test them.

1 A new hot drinks flask

a) What properties would the materials need to keep the drink hot? What other properties would help to make a flask that is easy to carry around? Is there anything else you would like the flask to do?

b) How would you test the insulating properties of your flask with a thermometer? What apparatus would you use? How would you ensure that your test is fair?

2 Some new ski clothing

a) What properties would the materials need to keep the wearer warm and dry? What other properties would help to make the clothes comfortable to wear? Is there anything else you would like the clothes to do?

b) How would you test the waterproof properties of your materials, and the speed at which they dry? How would you ensure that your test is fair?

A new diving suit

Scientists in the USA have developed a new diving suit for soldiers, using ducks as their inspiration. Ducks can swim under water but when they come to the surface, their feathers are dry. They can also keep warm in chilly waters. Scientists have developed a diving suit made with three layers. The first layer next to the skin reflects heat, to keep the diver warm. The second layer is made with a smart material – at low temperatures it is waterproof to keep the diver dry in the water. But at high temperatures, this material allows moisture to pass through. This means the fabric is breathable when the diver gets hot and begins to sweat. The outer layer is a streamlined shell to aid movement in the water. This lightweight suit can now be worn for jobs in and out of the water.

Glossary

adhesive A substance used to stick two surfaces together. Glue is an adhesive.

biodegradable When a material decays (breaks down) naturally.

boil When a liquid turns into a gas. Liquids boil when they are heated to a temperature called their boiling point.

burning When a material is heated and catches fire.

carbon fibre A strong, light thread. Carbon fibre is added to materials to make them stronger.

chemical reaction A reaction that causes a material to change permanently. Burning is an example of a chemical reaction.

compressed Squashed.

condensation When a gas turns into a liquid, usually because it has been cooled.

conductor A material that allows heat or electricity to pass through it.

dew Condensed water found on the ground in the early morning or evening.

diluted When the strength of a solution is made weaker by adding water or another liquid.

dissolve When a solid is absorbed by a liquid to make a solution.

elasticity When a material is stretched or squashed, but goes back to its original shape and size.

electricity A type of energy. We use electricity to power lights and to work machines.

evaporation When a liquid gradually turns into a gas. Liquids evaporate more quickly when they are heated.

expand To spread out and get bigger.

filtering A way of separating a solid from a liquid.

freeze When a liquid turns into a solid because it has been cooled.

gas A material that moves and expands to take the shape of the container it is in. Air is a gas.

global warming A rise in temperatures on Earth, partly caused by burning fossil fuels.

insulator A material that does not allow heat or electricity to pass through it.

irreversible A one-way change that cannot be undone. Burning is an irreversible change.

limestone A type of rock, often used as a building material.

liquid A material that flows and pours and takes the shape of the container it is in. Water is a liquid.

melt When a solid becomes a liquid. Many solids melt when they are heated.

mined Taken from the Earth. Some rocks are mined from beneath the Earth's surface.

mixture When different substances are mixed together.

moulded When the shape of a solid is changed by melting the solid and solidifying it into a new shape.

processed When a material is changed, by cutting or polishing, for example.

properties A description of what a material is like.

purity How pure or clean a substance is.

recycling When a material is reprocessed to make something new.

reversible A change that can be undone. Melting is a reversible change.

solid A material that is able to hold its shape. Wood is a solid.

solidifying When a liquid turns into a solid, often at room temperature.

solution A liquid containing a dissolved solid or gas.

substance A type of material.

temperature The amount of heat energy that a material has.

Venn diagram A diagram that divides items into sets. Items in overlapping areas have similar properties.

water vapour When liquid water turns into a gas.

Answers

Page 4: Example answer: Glass is made by heating sand, limestone and soda at high temperatures. Plastic is made by processing chemicals found in mineral oil. Paper is made from the wood from trees. The wood is chopped into tiny pieces and mixed with chemicals to form a pulp that is pressed and dried to make paper.

Page 5: Natural: twig (wood), water, rock. Man-made: paper, plastic bucket, glass. Natural but processed: ball of wool, wooden ruler, stone paperweight (polished).

Page 7: Example answers: Soft (pillow, good to sleep on). Smooth (bath, doesn't scratch the skin). Sharp (scissors, good for cutting). Shiny (mirror, can see your reflection). Transparent (window, lets light through). Light (lunch box, easy to carry). Absorbent (kitchen towel, wipes up spills). Waterproof (umbrella, keeps you dry). Other properties may include: hard, rough, strong, bendy, able to float.

Page 8: A solid's shape can be changed by cutting, bending or squeezing. Some solids can be melted and then moulded into different shapes.

Page 10: You should see a small amount of water rising in the cup. This is because air in the cup is compressed to fill a smaller space as the water pushes against it. When you tilt the cup, air from the cup escapes and rises as bubbles of gas. Water fills this space in the cup.

Page 11: Fizzy drinks are made by forcing carbon dioxide gas into a liquid and sealing the container. The gas dissolves in the liquid under pressure. When you open the container, bubbles of gas form and float to the surface with a 'fizzing' sound. If the drink is left without a lid for too long, all the gas escapes from the liquid and the drink goes 'flat'.

Helium is a gas that is lighter than air. When balloons are filled with helium gas they rise and 'float' in air.

Page 13: Some of the paint evaporates and turns into a gas. As it spreads across the room and reaches our nose, we begin to smell it.

Page 14: When you take a drinks can out of a cold fridge, water vapour in the air cools and condenses against the cold surface of the can.

Page 14: Drops of water form on the underside of the food wrap. The water has evaporated from the bowl and condensed on the cool surface of the food wrap. When the drops of water get too heavy, they fall into the glass. The water in the glass isn't salty because the salt doesn't evaporate. When ice cubes are added, the water droplets get bigger. They are largest nearer to the ice cubes because more water condenses in these cold areas.

Page 15: In the early morning or evening, the grass is cold because the Sun hasn't warmed the Earth's surface. During the day, the grass is warmer so condensation doesn't form.

Page 16: Metals and glass are heated to melt them. Liquid metal and liquid glass can be poured into moulds to change their shape. When the metal or glass cools, it solidifies and takes the shape of the mould.

Page 18: Toffee is made by heating sugar. The sugar begins to burn and turn brown as a chemical reaction takes place. Honeycomb (sponge toffee) is made by adding vinegar and baking soda to hot toffee. These substances react to produce carbon dioxide gas. Bubbles of gas are trapped in the sticky toffee as it cools.

Page 20: The white marks are salt. The sea water is salty. As your skin dries, the water evaporates, leaving salt crystals behind.

Page 21: Sieve the mixture into the bowl to separate the shells from the water, salt and sand. Filter the remaining liquid to separate the sand from the water and salt. Leave the salt water in a beaker for a few days until the water has evaporated. You can't extract the salt by filtering because it dissolves in the water.

Page 22: Example answers: tights, socks, springs, elastic band, hair band, elastic belt.

Page 23: For a fair comparison, the same number of marbles needs to be added to each sock (marbles of the same size). The socks also need to be measured from the same position each time.

Page 24: Metals are good conductors of heat. Radiators are filled with hot water. The metal conducts (passes) the heat to the air in the room to warm it up.

Page 25: For a fair comparison, the same amount of insulation is needed around the cups. The thermometer holes need to be the same size to stop additional heat escaping. Ideally, the thermometers should also be read at the same time.

Page 27: At most recycling centres you can recycle many materials such as paper, glass, cardboard and plastic.

Page 29: Example answers: Drinks flask (light, non-breakable, able to see how much drink is inside, easily washed, insulating to retain heat). Ski clothing (waterproof, insulating, breathable, easy to dry, bright for visibility, soft, flexible).

Websites

http://www.strangematterexhibit.com/
Find out more about everyday materials.

http://www.bbc.co.uk/schools/ks2bitesize/science/materials.shtml
Learn about materials and try out some fun activities.

http://www.crickweb.co.uk/assets/resources/flash.php?&file=materials2d
Test your knowledge about materials.

http://www.sciencenewsforkids.org
Keep up with the latest news about materials.

Index

The Science Detective Investigates

Contents of titles in the series:

WAYLAND